SCIENCE
Q & A

Light

Gina L. Hamilton

LIGHTBOX

Go to
www.openlightbox.com
and enter this book's
unique code.

ACCESS CODE

LBE46788

Lightbox is an all-inclusive digital solution for the teaching and learning of curriculum topics in an original, groundbreaking way. Lightbox is based on National Curriculum Standards.

STANDARD FEATURES OF LIGHTBOX

 AUDIO High-quality narration using text-to-speech system

 ACTIVITIES Printable PDFs that can be emailed and graded

 SLIDESHOWS Pictorial overviews of key concepts

VIDEOS Embedded high-definition video clips

WEBLINKS Curated links to external, child-safe resources

 TRANSPARENCIES Step-by-step layering of maps, diagrams, charts, and timelines

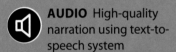

 INTERACTIVE MAPS Interactive maps and aerial satellite imagery

QUIZZES Ten multiple choice questions that are automatically graded and emailed for teacher assessment

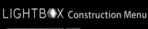

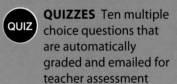

 KEY WORDS Matching key concepts to their definitions

CONTENTS

What Is Light?

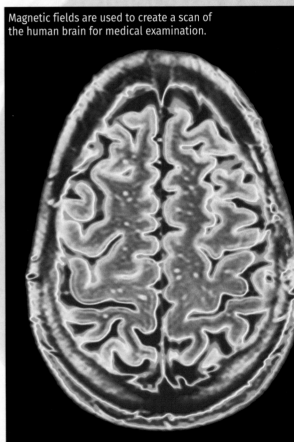

Magnetic fields are used to create a scan of the human brain for medical examination.

Light is a form of **electromagnetic energy**. This energy is made up of changing electric and **magnetic fields**. Light may be visible or invisible. Different types of visible and invisible light make up the **electromagnetic spectrum**. The tiny part of the electromagnetic spectrum that humans can see is called the visible light spectrum. Just as waves in water ripple out from their source, light can be thought of as a wave that spreads out in all directions from a light source, such as the Sun or a light bulb. Light gets dimmer as it moves farther away from its source. Light waves radiate, or travel, along straight lines called rays. This process is called radiation. The light that the human eye can see is only one type of radiation.

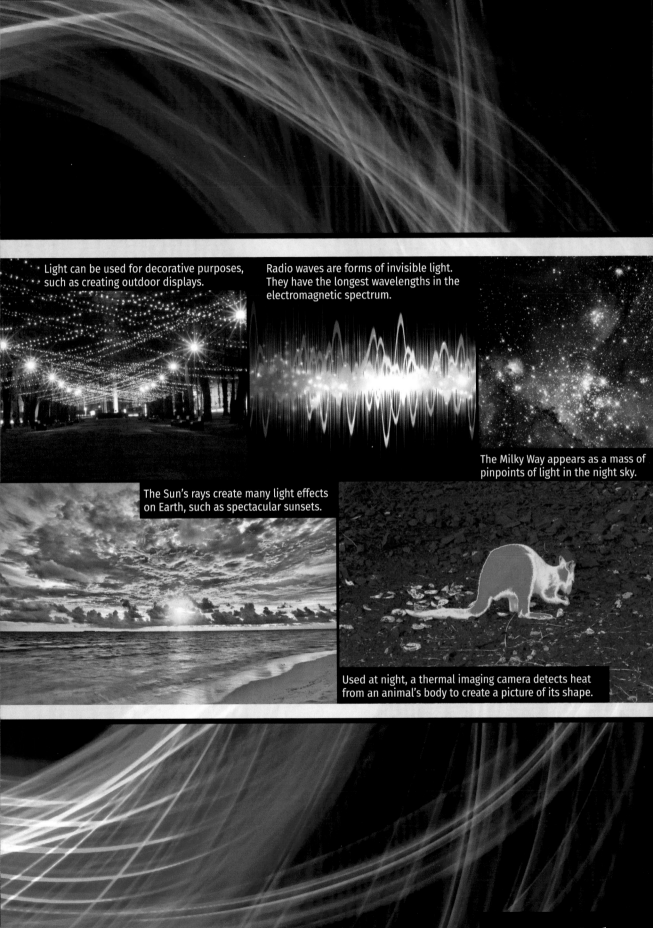

Light can be used for decorative purposes, such as creating outdoor displays.

Radio waves are forms of invisible light. They have the longest wavelengths in the electromagnetic spectrum.

The Milky Way appears as a mass of pinpoints of light in the night sky.

The Sun's rays create many light effects on Earth, such as spectacular sunsets.

Used at night, a thermal imaging camera detects heat from an animal's body to create a picture of its shape.

Why Does Light Matter?

What would life on Earth be like without light and energy from the Sun? Light is essential to our existence. Without it, people would be unable to see anything, and most living things on Earth would not survive. Light from the Sun provides energy for life on Earth.

Most life on Earth depends on light and the relationship between plants and light. Plants turn sunlight into food through a process called **photosynthesis**. Photosynthesis is the first step in the food chain that connects almost all living things. Plants are a basic source of food because almost all animals eat plants, the products of plants, or other animals that eat plants. For example, cattle and other farm animals graze in fields of grass. Many people then eat meat and drink milk that comes from these animals.

Giraffes eat the leaves of bushes and trees. On average, a giraffe will eat 66 pounds (30 kilograms) of food per day.

In addition to the role that it plays in photosynthesis, another important aspect of sunlight is the warmth that it provides. The Sun's energy keeps Earth at a temperature suitable for living things to survive.

Your Challenge!

Find two small plants. Place one plant in a warm, sunny area. Place the other in a warm, shaded area where it will receive no direct sunlight. Give each plant 1 tablespoon of water every day at the same time. Record your observations daily for six days. What did you observe about the two plants? What conclusions can you make?

1814 The world's first gas-powered street lights were erected in London, England.

1899 Serbian-American scientist Nikola Tesla succeeded in transmitting electricity without wires to illuminate 200 lamps in Colorado Springs.

Light through History

3000BC

Candles were invented by the Ancient Egyptians. The candles were made from beeswax.

1879

Thomas Alva Edison invented the electric light bulb.

1790 The Betty lamp was invented in Europe. The lamp burned fish oil or animal fats to produce light.

1898 Neon gas was discovered. Neon tube lighting was then invented in 1911. Neon lighting later became popular for creating advertising signs.

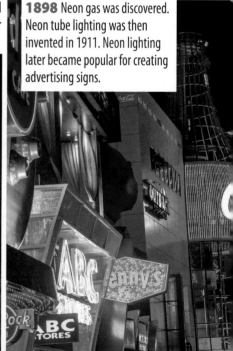

1962 Laser beams were used to create the first three-dimensional image, known as a "hologram."

2004 The Keck Observatory in Hawaii used the first laser guidance system on a large telescope. This allowed astronomers to take highly detailed photographs of the solar system.

Dutch electronics company Philips invented a fluorescent light bulb, with a 60,000-hour lifespan.

1900

1991

FUTURE

1994 The first high-brightness light-emitting diode (LED) lights were developed.

2016 A project called Breakthrough Starshot was launched. It aims to use laser beams on Earth to send tiny vehicles into space to seek out new forms of life.

DEPARTURES

	Destination	Flight	
19:30	SAO PAULO	R4	450
19:30	BOGOTA	EB	718
19:45	RIO DE JANEIRO	DN	004
19:40	SANTIAGO	OD	715
19:50	LIMA	NP	689
20:05	QUITO	UC	120
20:10	CARACAS	EB	343
20:20	BUENOS AIRES	R4	458
20:45	MONTEVIDEO	NP	197

How Much Sunlight Does Earth Receive?

Every day, the Sun provides Earth with large amounts of energy in the form of sunlight. Scientists have made great strides in developing ways to capture this light. They have invented solar cells that collect light from the Sun to power equipment, light homes, and heat up water.

Solar cells work by changing light energy into electricity. Energy converted from sunlight is one of the most promising sources of **alternative energy**. Many scientists believe that solar energy will help to reduce pollution and the reliance upon fossil fuels, such as oil and gas. Solar cells operate best where there is bright sunshine, but can still produce electricity in cloudy conditions.

Scientists make a distinction between bright sunshine and lower levels of daylight. Bright sunshine reaches Earth's surface when the skies are clear. Different regions of Earth receive different amounts of bright sunshine, depending on the climate.

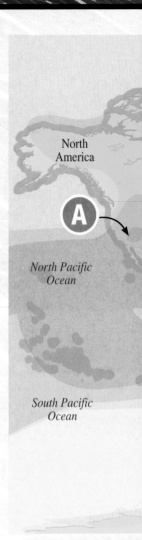

North America

A

North Pacific Ocean

South Pacific Ocean

A
North America
Yuma, Arizona.
4,015 hours of bright sunshine per year

B
South America
Santiago, Chile.
3,783 hours of bright sunshine per year

C
Europe
Valletta, Malta.
3,054 hours of bright sunshine per year

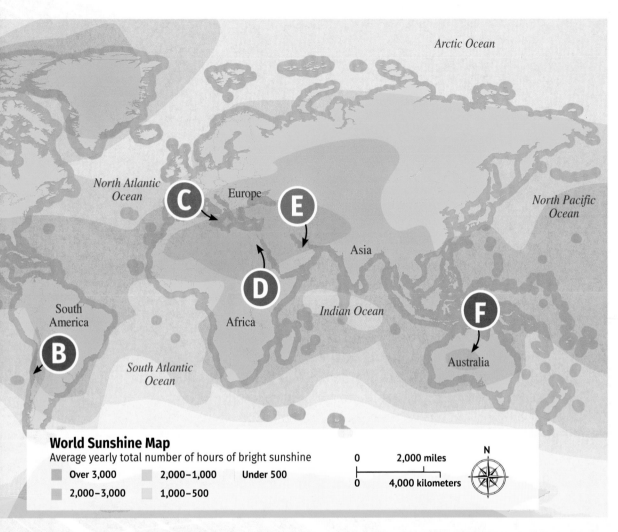

Arctic Ocean

North Atlantic
Ocean

Europe

North Pacific
Ocean

Asia

C

E

D

Africa

Indian Ocean

F

South
America

South Atlantic
Ocean

Australia

B

World Sunshine Map
Average yearly total number of hours of bright sunshine

- Over 3,000
- 2,000–3,000
- 2,000–1,000
- 1,000–500
- Under 500

0   2,000 miles
0   4,000 kilometers

N

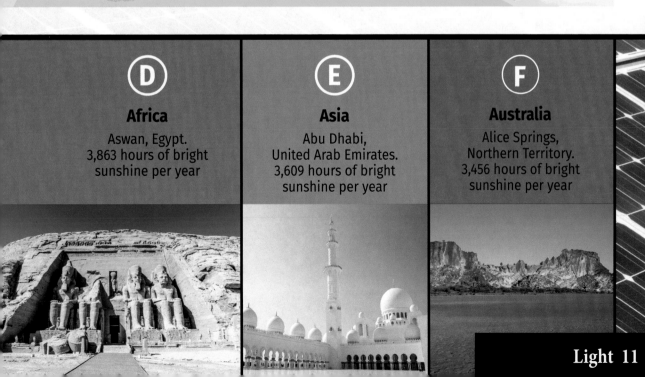

D

Africa

Aswan, Egypt.
3,863 hours of bright
sunshine per year

E

Asia

Abu Dhabi,
United Arab Emirates.
3,609 hours of bright
sunshine per year

F

Australia

Alice Springs,
Northern Territory.
3,456 hours of bright
sunshine per year

Is Light a Particle or a Wave?

Light acts like a wave. It also, however, acts like a particle. For hundreds of years, scientists have debated whether light is a particle or a wave. Some ancient Greek scientists believed light was made of a stream of very small **particles**.

Albert Einstein was awarded the Nobel Prize for his discovery that light consists of particles called photons.

Sir Isaac Newton believed in this particle theory of light in the early 1700s. At about the same time, Christiaan Huygens came up with a different theory. He thought light acted more like a wave than a stream of particles. He saw that light spread out once it passed through a small opening. If the particle theory were correct, the light would remain in a beam as it passed straight through the opening. Over the years, scientists began to agree with Huygens' theory. By the nineteenth century, the wave theory had become the most accepted theory of the behavior of light.

In 1905, physicist Albert Einstein developed a different theory of light. He believed that light has some characteristics of a particle and some characteristics of a wave. This idea is known as the wave-particle duality theory. Einstein observed that, under certain circumstances, light waves behaved as though they were made of individual energy packets. These energy packets are known as **photons**. Einstein's revolutionary ideas surprised scientists all over the world.

Your Challenge!

Use a flashlight to investigate the way light travels. Cut a series of slits close together through a piece of paper or index card. Make the slits large enough for light to pass through. Then, in a dim or dark room, hold the flashlight so it shines through the slits in the paper. You may need to adjust the paper or index card and the flashlight in order to see the light effect. What do you observe?

At What Speed Does Light Travel?

Light is the fastest thing in the universe. It can travel around Earth seven times in one second. The Sun is 93 million miles (150 million kilometers) from Earth. It takes approximately eight minutes for light to travel from the Sun to Earth. In a **vacuum**, light travels at about 186,000 miles (299,792 km) per second.

Light travels at
671 million miles
(1,080 million km)
per hour.

The speed of light was discovered by Armand Fizeau in 1849. He timed it using a spinning cogwheel—a special wheel that had small teeth, or gaps, evenly spaced around its edge—and a mirror set up about 5.4 miles (8.6 km) away. Light passed through one gap between the teeth of the wheel as it traveled to the mirror and, if the wheel was turning quickly enough, through a neighboring gap on the way back. By knowing the speed of the wheel, Fizeau could calculate the speed of light as it traveled to the mirror and back.

If an object traveled continuously at the speed of light for one year, it would cover a distance of more than 5.8 trillion miles (9.5 trillion km).

It takes **100,000 YEARS** for light to travel from one end of the galaxy to the other.

Later, scientists wondered if light projected from the front of a moving train traveled at the speed of light plus the speed of the train. Physicist Albert Einstein learned the answer in 1905. Einstein discovered that the speed of light remains constant. It never changes, no matter the speed of the train. This is called Einstein's special theory of relativity.

A simple explanation of the special theory of relativity is that nothing travels faster than the speed of light. As an object approaches the speed of light, it becomes heavier and needs more energy to make it move. This means that even a tiny object would become infinitely heavy at the speed of light, and it would need an infinite supply of energy to keep it moving at that speed.

What Is Invisible Light?

Visible light, the small group of light waves in the middle of the electromagnetic spectrum, is the type of light that is most familiar to humans. Objects such as the Sun and stars emit visible light, as well as invisible light. Invisible light is different forms of radiation on either side of the visible light spectrum. Each type of light has a different **wavelength**. The wavelengths on either side of the visible light spectrum are important. These wavelengths are too long or too short to be detected by the human eye. Even though most people never think about it, invisible light is everywhere.

Radio waves are forms of invisible light. They have very long wavelengths and are at one end of the electromagnetic spectrum. They are used to send radio and television signals. Microwaves are also at this end of the spectrum and are often used for cooking and in radar technology. X-rays and gamma rays have very short wavelengths and are at the opposite end of the spectrum. These high-frequency waves are used mostly in medicine. For example, gamma rays are used to destroy cancer cells and sterilize, or clean, hospital equipment.

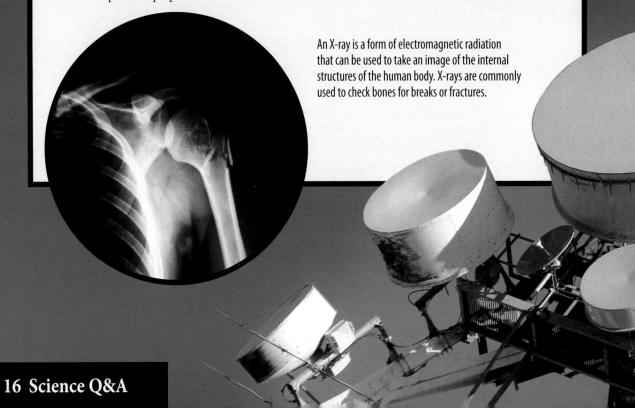

An X-ray is a form of electromagnetic radiation that can be used to take an image of the internal structures of the human body. X-rays are commonly used to check bones for breaks or fractures.

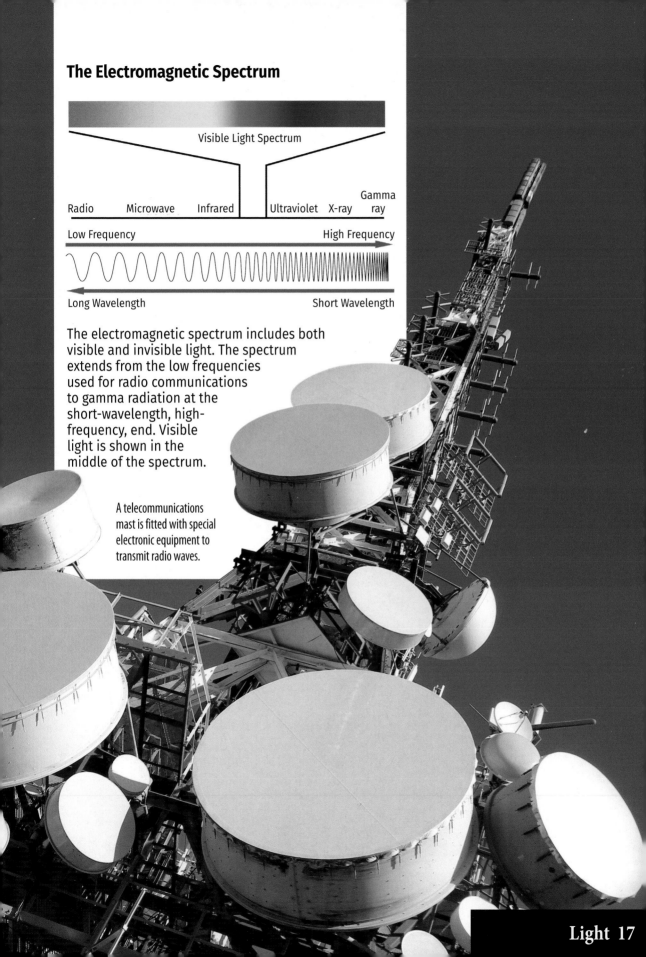

The Electromagnetic Spectrum

Visible Light Spectrum

| Radio | Microwave | Infrared | | Ultraviolet | X-ray | Gamma ray |

Low Frequency → High Frequency

Long Wavelength ← Short Wavelength

The electromagnetic spectrum includes both visible and invisible light. The spectrum extends from the low frequencies used for radio communications to gamma radiation at the short-wavelength, high-frequency, end. Visible light is shown in the middle of the spectrum.

A telecommunications mast is fitted with special electronic equipment to transmit radio waves.

What is Infrared Light?

I nfrared light is apart of the invisible light spectrum. It is found next to the red end of the visible light spectrum. Sir Frederick William Herschel discovered infrared light in 1800. He was the first person to discover that there were forms of light that human eyes cannot see.

Anything with a temperature gives off radiation. This radiation is in the form of infrared light. Infrared light cannot be seen by people, but it can be measured, and even felt. Very hot objects, such as hot coals, may not give off light that can be seen by the human eye, but they give off infrared radiation that can be felt as heat. Even very cold objects, such as ice cubes, give off some infrared radiation.

Most people use infrared radiation every day without even knowing it. Television and stereo remote controls use infrared technology. When a button is pressed on the remote control, it sends an infrared light signal. The television or stereo receives the signal and responds to it.

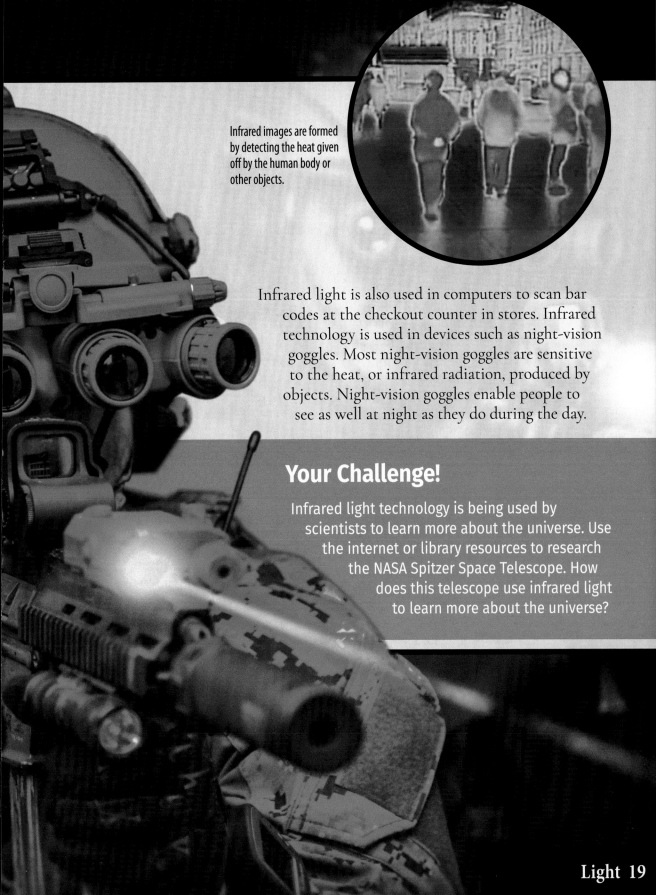

Infrared images are formed by detecting the heat given off by the human body or other objects.

Infrared light is also used in computers to scan bar codes at the checkout counter in stores. Infrared technology is used in devices such as night-vision goggles. Most night-vision goggles are sensitive to the heat, or infrared radiation, produced by objects. Night-vision goggles enable people to see as well at night as they do during the day.

Your Challenge!

Infrared light technology is being used by scientists to learn more about the universe. Use the internet or library resources to research the NASA Spitzer Space Telescope. How does this telescope use infrared light to learn more about the universe?

What Are Laser Beams?

Most light spreads out into a wide circle, strongest in the center of the circle and weakest at the outer edge of the circle. Light that is seen from the Sun, a lamp, or a flashlight is a mixture of many different wavelengths that, when mixed together, appear to be white. The wavelengths are all different, so they are incoherent, or out of step with one another. This spreads the light waves in all directions. Laser light does not do this. It is all one wavelength and color, and all the waves are in step with one another. They are coherent.

The word "**Laser**" stands for **L**ight **A**mplification by the **S**timulated **E**mission of **R**adiation.

Laser beams are created when atoms that are contained within a tube are stirred up by electricity. When the atoms are treated this way, they produce coherent light waves. Focusing these light waves with mirrors and **lenses** produces an intense beam of coherent light. This is a laser beam.

Lasers always travel in a straight line without spreading out. This means that they can be used in construction and other fields where a straight, accurate line is necessary for precise measurements. Laser light is also a powerful tool in medicine, because lasers can cut with precision. Laser light can be used to correct vision, drill teeth, and remove diseased tissue during surgery.

A **laser beam** can be **hotter** than the surface of the Sun, which is **10,000°** Fahrenheit (5,600° Celsius).

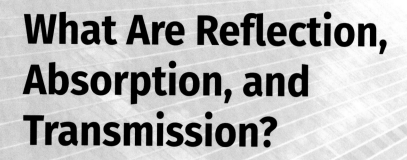

What Are Reflection, Absorption, and Transmission?

What happens to light when it hits an object? Does it bounce off, go inside the object, or go straight through? Do different objects cause light to react in different ways?

Just as a ball bounces off a wall or floor, light bounces off some objects. This is called reflection. Most uneven surfaces, such as clothing, paper, and skin, reflect light and scatter it in many directions. For example, if a person looks at his or her reflection in a piece of unpolished metal, which has an uneven surface, the image appears blurry or distorted.

Some objects are better reflectors than other objects. Smooth and shiny surfaces reflect light very well. Perfectly smooth surfaces allow light to reflect evenly. Mirrors are nearly perfect reflectors. This means that almost all of the light that shines on a mirror reflects off it at the same angle.

Some surfaces take in light. This is called absorption. When light is absorbed, it is taken in and is unable to reflect. Think about a black car and a white car on a hot, sunny day. Since black objects absorb light, the surface of the black car will be much hotter to the touch than the white car, which reflects light.

During transmission, light is not reflected or absorbed. When light is transmitted, it passes right through a surface or object. Clear glass is an example of a surface that transmits light.

Why Do Only Some Objects Cast Shadows?

Why does a tree cast a shadow, but a glass door does not? There are some objects that people can see through and some that they cannot. A transparent object allows most of the light it comes into contact with to pass through it. A goldfish bowl is transparent.

A translucent object allows some light to pass through it, but it does not allow enough light through so that objects can be seen clearly on the other side of it. The light changes direction many times and is scattered as it passes through. Therefore, objects cannot be seen clearly through translucent materials and appear fuzzy. Frosted glass and some plastics are examples of translucent materials.

Opaque objects, such as a solid structure, cast shadows. A brick wall is opaque. A person or a statue is also opaque. A shadow is the area behind an opaque object where light cannot reach. A shadow shows the outline of the object that is blocking the light. An opaque object, such as a basketball, casts a shadow. The size of the shadow that the ball casts shrinks as the ball is moved away from a light source. For example, if the ball is near a light bulb, it casts a large shadow on the wall. If the ball is placed farther away from the bulb, it blocks less light, so the shadow is smaller.

A sundial is a clock that uses the position of the Sun to indicate the time. A pointer in the center casts a shadow onto a plate that is marked with the hours of the day.

What is Fiber Optics?

Fiber optic cables are used to move information at high speeds across long distances. These cables carry light instead of electricity. They are not affected by temperature changes, rain, or virtually any other environmental condition. Fiber optic cables consist of many long fiber optic strands made of glass or plastic. Each fiber optic strand is as thin as a strand of human hair. Hundreds of thousands of fiber optic strands are placed together in bundles to form fiber optic cables. Sounds, images, and other information are changed into pulses of light. These pulses of light then travel through the fiber optic cables.

A fiber optic strand, called the core, is covered by a transparent coating. This coating is called cladding. Light usually transmits through transparent surfaces, but the cladding is designed to block the light, sending it back into the core.

A typical fiber optic cable can carry up to

250 TELEVISION CHANNELS.

The light pulses travel through the core by bouncing back and forth on the cladding. As long as the curves in the fiber optic cable are never too extreme, the light pulses always hit the cladding at an angle, allowing them to be reflected forward through the core. Using cladding on the core allows fiber optic cables to be installed around corners, through walls, and underground.

Fiber optic cables are widely used for carrying internet, TV, and telephone signals. Fiber optic cables can carry more information more securely than metal wires of the same diameter.

The light pulses in a fiber optic cable travel at **60,000** MILES (96,500 km) per second.

What Are the Primary Colors?

A primary color is any of a group of colors from which all other colors can be made by mixing. A primary color cannot be produced by mixing two colors together, and the primary colors are combined to produce all the other colors, or hues. Colors can be mixed either by combining colors of light or by combining colored **pigments**.

The primary colors of light are red, blue, and green. They can be added together to produce every color. Look closely at a television screen while it is on. There are tiny dots of red, blue, and green light. In various combinations, these colors produce all the colors seen on the screen.

The primary colors of pigments are red, yellow, and blue. These are the colors used to make dyes, inks, and paints. Pigments absorb some colors of light and bounce others back to the eye. Absorbed colors cannot be seen. Only the colors that bounce back are visible.

The three primary colors of light are blue, green, and red. When all three colors are mixed together in equal amounts, the resulting light is white.

Your Challenge!

Combinations of primary colors create all other colors. Cut a piece of red cellophane, or plastic wrap, large enough to cover the end of a flashlight. Use an elastic band or tape to hold it in place. Do the same with blue and green cellophane using two additional flashlights. Shine any two flashlights at the same spot on a white piece of paper. Record the color that is created. Do this with different combinations of flashlights. Are you surprised at some of the results?

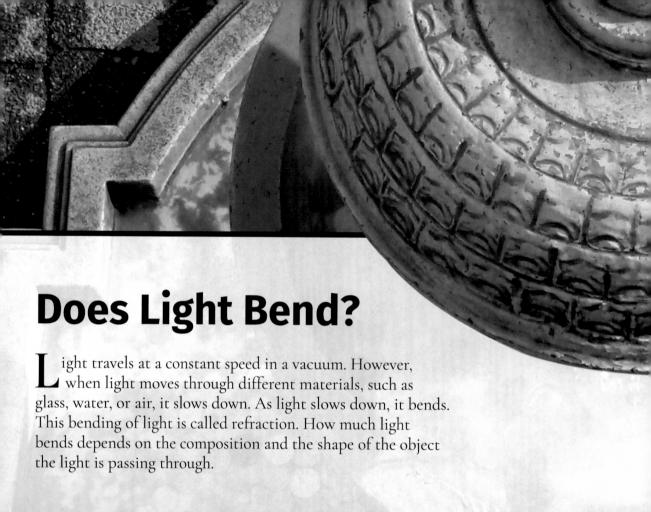

Does Light Bend?

Light travels at a constant speed in a vacuum. However, when light moves through different materials, such as glass, water, or air, it slows down. As light slows down, it bends. This bending of light is called refraction. How much light bends depends on the composition and the shape of the object the light is passing through.

Refraction can play tricks on the eyes. If a person reaches for a coin at the bottom of a deep fountain, he or she will discover that it is in a different place than it appeared to be. This is because the light that reaches the eyes from the coin has been refracted.

As light passes from air to water, its speed slows down and the light rays are bent, distorting how objects appear.

How Do Human Beings See Light?

Light enters the human eye through a clear covering called the cornea, which passes light through the lens. The lens focuses the light on the retina, a layer of light-sensitive cells at the back of the eye. The retina changes the light into signals, which travel through the optic nerve to the brain. The brain reads the signals as an image. For people with perfect eyesight, the images are focused directly on the retina.

Most people do not have perfect eyesight. Some people can see objects close up fairly well, but things that are farther away look fuzzy and blurry. This is caused by the **focal point** of the light occurring in the eye before it reaches the retina. This condition is called nearsightedness. Other people see distant objects fairly well, but things seen up close are blurry. This kind of vision is called farsightedness. In this case, the image has a focal point beyond the retina.

Eyeglasses and contact lenses are used to correct nearsightedness and farsightedness. They contain lenses that bend, or refract, the light to focus properly on the retina. Lenses allow people who are nearsighted and farsighted to see as well as people with perfect vision. Eyeglasses and contact lenses are used to correct nearsightedness and farsightedness. They contain lenses that bend the light to focus properly on the retina. Lenses allow people who are nearsighted and farsighted to see as well as people with perfect vision.

Your Challenge!

When light enters the eye, an upside-down image is formed on the retina. Research pinhole cameras, then try to build your own. Using the results of your research, explain why the image appears upside down, or inverted, inside the pinhole camera.

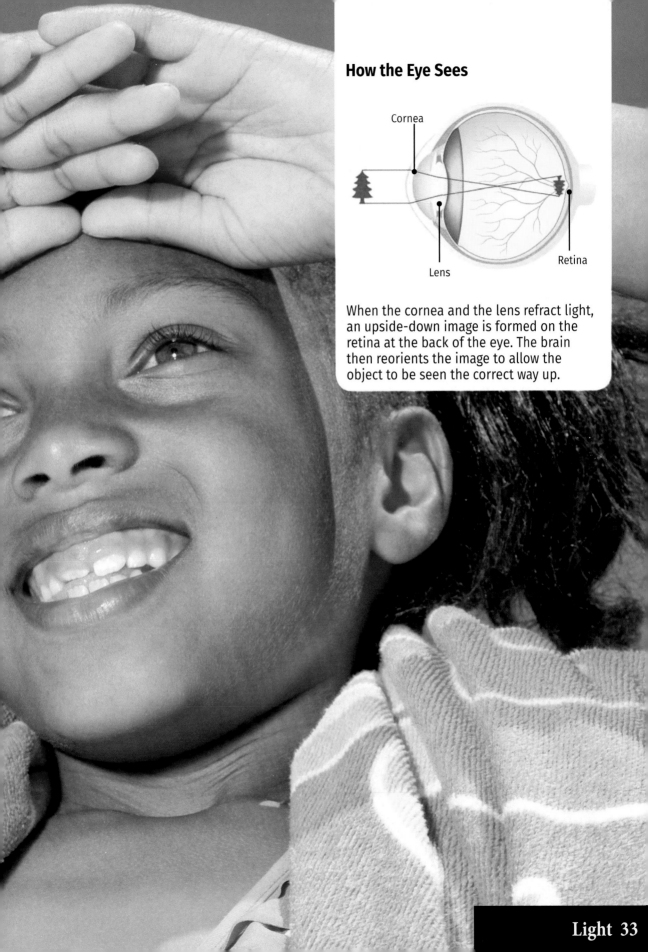

How the Eye Sees

Cornea

Lens

Retina

When the cornea and the lens refract light, an upside-down image is formed on the retina at the back of the eye. The brain then reorients the image to allow the object to be seen the correct way up.

How Does the Human Eye Adjust To Light?

The human eye is constantly adjusting itself to different levels of light. Usually, the eye adjusts itself without the person even thinking about it. These are involuntary responses by the body. Sometimes, people adjust the amount of light entering the eye using a voluntary response. Squinting, for example, is one way that people use a voluntary response to adjust the amount of light entering the eye.

In order to see things clearly, the amount of light coming from an object must be controlled. In near-darkness, the pupil expands. This allows more light into the eye. In bright sunlight, the pupil contracts, or gets smaller, protecting the eye from being flooded with light. This process takes a moment or two, which is why it takes a while for a person's eyes to adjust to light and dark.

When a person sustains a head injury, the pupils can stop functioning properly. Pupils that are unequal in size are a sign of a brain injury. Another sign is if the pupils are unresponsive to light. When a light is shined directly onto a person's eyeball and the pupils do not change, the person may have received a brain injury.

When watching a 3D movie, people wear special eyeglasses. These glasses adjust the eyes to different types of light to create three-dimensional images on the screen.

How Do Human Beings See Color?

Humans use color for many different purposes. How do people tell the difference between all of the different colors in the spectrum? For example, how do people tell the difference between the red signal and the green signal on a traffic light? The retina contains nerve cells called rods and cones. Rods and cones send messages to a type of cell called a ganglion. Ganglion cells then transmit the messages to the brain, where they are turned into images.

There are three types of cones. Red cones are sensitive to red light, green cones are sensitive to green light, and blue cones are sensitive to blue light. Together, these three types of cones allow people to see a large range of different colors.

Every time a person sees a colored object, some of the light is being reflected, while some is absorbed. The color that is reflected to the eye gives the object its color. The rest of the wavelengths are absorbed into the object. When a person sees something that is green, it means that only the green wavelengths of light are reflected back to their eyes. In plants, a pigment called chlorophyll absorbs all the colors of light except green. This is why the leaves of trees appear green.

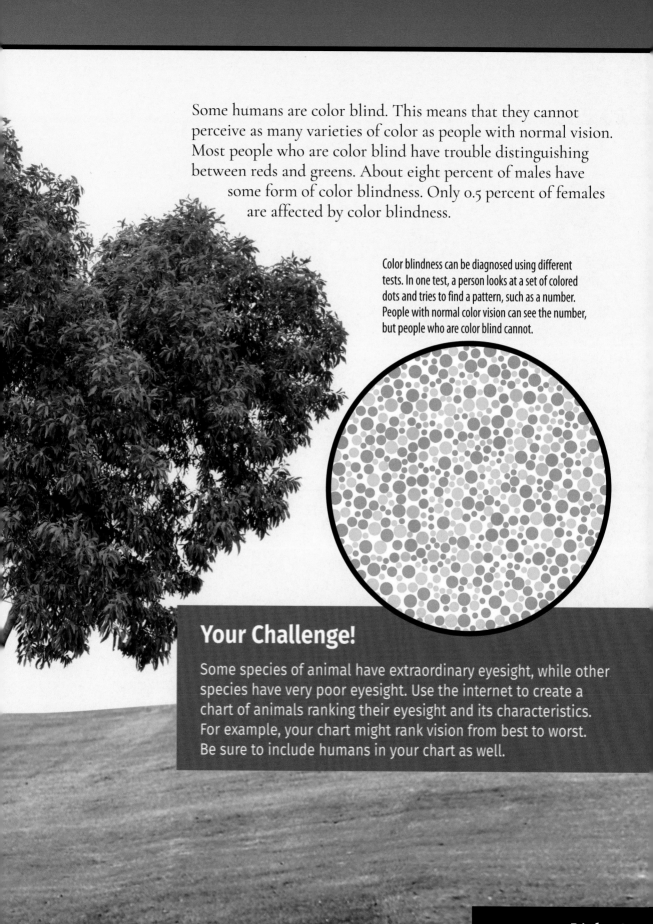

Some humans are color blind. This means that they cannot perceive as many varieties of color as people with normal vision. Most people who are color blind have trouble distinguishing between reds and greens. About eight percent of males have some form of color blindness. Only 0.5 percent of females are affected by color blindness.

Color blindness can be diagnosed using different tests. In one test, a person looks at a set of colored dots and tries to find a pattern, such as a number. People with normal color vision can see the number, but people who are color blind cannot.

Your Challenge!

Some species of animal have extraordinary eyesight, while other species have very poor eyesight. Use the internet to create a chart of animals ranking their eyesight and its characteristics. For example, your chart might rank vision from best to worst. Be sure to include humans in your chart as well.

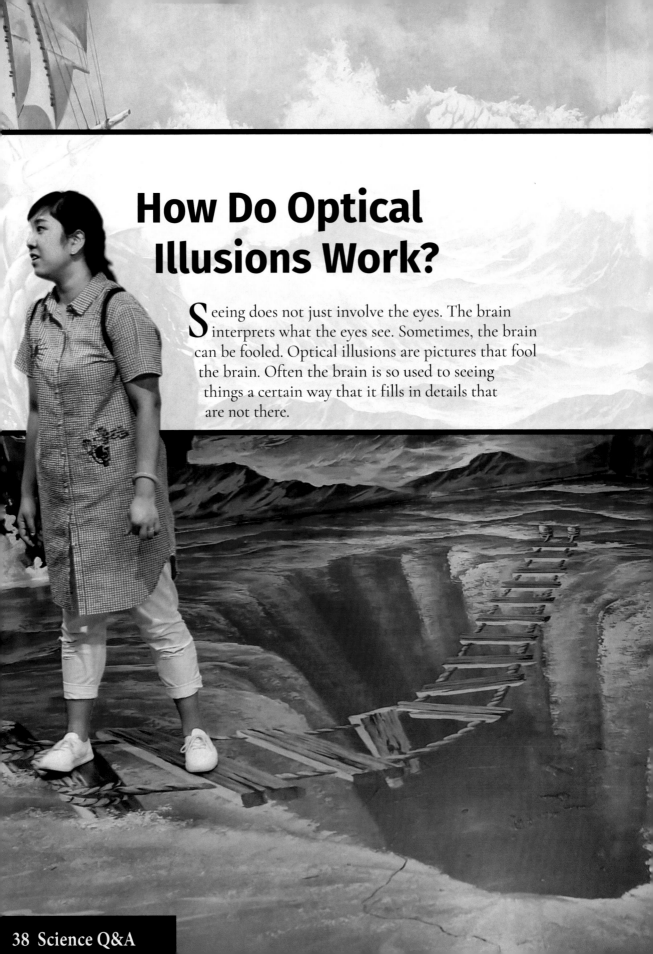

How Do Optical Illusions Work?

Seeing does not just involve the eyes. The brain interprets what the eyes see. Sometimes, the brain can be fooled. Optical illusions are pictures that fool the brain. Often the brain is so used to seeing things a certain way that it fills in details that are not there.

The information that is gathered by the eyes is processed by the brain, and the brain interprets what the eyes see and arranges it into a pattern that makes sense. Optical illusions happen because color, light, and patterns create images that can be deceptive or misleading to the brain. Optical illusions trick the brain into perceiving something differently to the way it actually is, so what a person sees does not match what is really there.

A mirage is an optical illusion that can show water on a highway, when, in fact, there is none.

Your Challenge!

Use the internet or other resources to find optical illusions. Develop a procedure to interview several people and test them with the optical illusions you found. Record your results and see if you can identify any patterns.

How Do Lenses Help People To See?

People who wear eyeglasses or contact lenses, or anyone who uses a magnifying glass, puts the science of lenses to use. A lens is a curved piece of transparent material used to refract light. Lenses are usually made of glass or plastic, and they have two basic shapes—concave, which curves inwards, and convex, which curves outwards.

A concave lens is thicker around its edges than it is in the middle. These lenses make an object appear smaller and farther away. Concave lenses are sometimes used in cameras. A convex lens is thicker in the middle than at the edges. A convex lens helps to bring an image into focus. It can also be used to make an object appear larger than it actually is. Convex lenses are used in cameras, binoculars, microscopes, magnifying glasses, and telescopes.

Microscopes use convex lenses to magnify objects. Even human skin can be examined in great detail.

Putting It All Together

Light is a form of energy. The Sun is the main source of light in the solar system. The Sun's light is essential to life on Earth. Light has many properties. It can be reflected, transmitted, absorbed, and refracted. Some light is visible to the human eye. This is the visible light spectrum. Colors in the visible light spectrum include red, orange, yellow, green, blue, indigo, and violet. Other light is invisible to the human eye. The invisible light wavelengths include radio, microwave, infrared, ultraviolet, X-ray, and gamma ray. Together, all of the different types of light make up the electromagnetic spectrum. Light is the fastest thing in the universe. The speed of light is about 186,000 miles (299,792 km) per second.

Since ancient times, people have been curious about light: how it acts, its effects, and its many uses. People use light and its energy every day. New ways to use light in technology, in the home, at work, in medicine, and as an alternative energy resource are discovered and developed all the time. Human beings are realizing, now more than ever before, how important it is to understand the relationship that exists between light and the planet Earth that is home to so many different species and organisms.

Scientists conduct experiments to understand more about light and its properties. This, in turn, helps us to understand how light affects our lives, our planet, and our understanding of the universe. The more that we learn about light, the more questions and answers we discover.

The Northern Lights, or *aurora borealis*, appear in the night sky in the Arctic region of Earth. They form when fast-moving, electrically charged solar particles collide with air molecules in Earth's atmosphere.

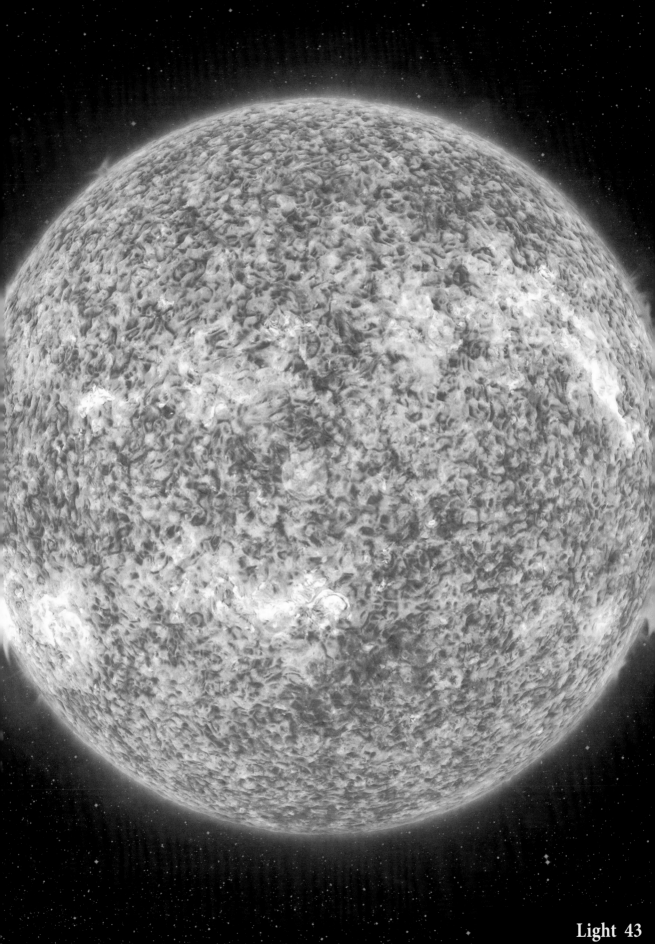

Light 43

Careers

Astronomer

Astronomers spend their time looking at the skies. They study galaxies, the Moon, planets, stars, and the Sun to learn more about the universe. Astronomers work in many different places. They can plan space flights or study the information that is gathered by satellites, telescopes, and observatories. Many people and organizations, such as the National Aeronautics and Space Administration (NASA), weather forecasters, museums, planetariums, and universities, rely on astronomers for information. Becoming an astronomer takes hard work. Most astronomers have a doctoral degree in astronomy. This can take as long as eight years to achieve.

Optometrist

Optometrists, or eye doctors, help people with eye problems. They perform tests on a patient's eyes to find out what prescription they need in order to see better. Optometrists can test for illnesses in the eyes that may cause blindness or other vision problems. When the patient's prescription is determined, the optometrist can help pick

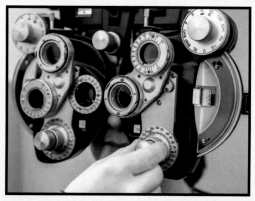

out a correct pair of glasses or contact lenses. To be an optometrist, a person must study for many years. Most optometrists begin with a four-year science degree from a university. Once completed, they then take a four-year training program at a school of optometry.

Young Scientists at Work

Test Your Knowledge:

Test your knowledge of light with these questions and activities. You can probably answer the questions using only this book, your own experiences, and your common sense.

Fact:

The length of an object's shadow depends on the location of the light source in relation to the object.

Test:

Place an object, such as a glass of milk, on a table. Turn a flashlight on, and move it to different positions in front of the glass of milk while observing the length of the shadow that appears behind the glass of milk.

Predict:

Where is the flashlight when you make the following shadows?

      a. the longest

      b. the shortest

      c. the widest

      d. the narrowest

Answers: a. When the light is held low, close to the table. **b.** When the light is held over the top of the glass **c.** When the light is held very close to the glass **d.** When the light is held far away from the glass.

1 In 1905, which scientist observed light energy packets known as photons?

2 What is the speed of light in a vacuum?

3 What is the name of the pigment that gives leaves their green color?

4 What are the three reactions light can have when it hits an object?

5 What are the two basic shapes of lenses?

Quiz

Now that you have read the book, test your knowledge by answering these questions. The answers are provided below.

ANSWER KEY
1. Albert Einstein **2.** 186,000 miles (299,792 kilometers) per second **3.** Chlorophyll **4.** Reflection, absorption, and transmission **5.** Concave and convex **6.** Glass and plastic **7.** Red, blue, and green **8.** Refraction **9.** The eyes **10.** Optical illusions

6 Fiber optic cables can be made from which two materials?

7 What are the three primary colors of light?

8 By what term is the bending of light known?

9 Rods and cones are part of which organs of the human body?

10 Images that are deceptive or misleading to the brain are known as what?

Key Words

alternative energy: energy sources that are not based on burning fossil fuels

electromagnetic energy: energy in the form of electrical or magnetic waves that travel at the speed of light

electromagnetic spectrum: the different wavelengths of light energy, including visible light

focal point: the spot where light rays meet after they are bent by a lens

lenses: clear objects that bend and focus light

magnetic fields: the space around a magnet that attracts metals

particles: very small bits of matter, such as atoms

photons: particles of light

photosynthesis: the process plants use to make their own food from carbon dioxide, water, and sunlight

pigments: a substance that gives an object its color by absorbing and reflecting different types of light

vacuum: a space with no matter in it, not even air

wavelength: the distance between the top of one wave and the top of the wave behind it

Index

LIGHTB◆X

➕ SUPPLEMENTARY RESOURCES

Click on the plus icon ➕ found in the bottom left corner of each spread to open additional teacher resources.

- Download and print the book's quizzes and activities
- Access curriculum correlations
- Explore additional web applications that enhance the Lightbox experience

LIGHTBOX DIGITAL TITLES
Packed full of integrated media

VIDEOS

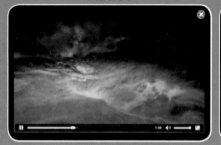

INTERACTIVE MAPS

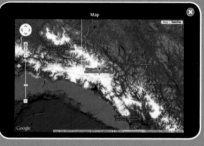

WEBLINKS

SLIDESHOWS

QUIZZES

OPTIMIZED FOR
- ✓ TABLETS
- ✓ WHITEBOARDS
- ✓ COMPUTERS
- ✓ AND MUCH MORE!

Published by Smartbook Media Inc.
350 5th Avenue, 59th Floor New York, NY 10118
Website: www.openlightbox.com

Project Coordinator: Heather Kissock
Art Director: Terry Paulhus

Library of Congress Cataloging-in-Publication Data

Names: Hamilton, Gina L., author.
Title: Light / Gina L. Hamilton.
Description: New York, NY : Smartbook Media Inc., [2016] | Series: Science Q & A | Includes index.
Identifiers: LCCN 2016051371 (print) | LCCN 2016051777 (ebook) | ISBN 9781510522398 (hard cover : alk. paper)| ISBN 9781510522404 (Multi-user ebk.)
Subjects: LCSH: Light--Juvenile literature. | Optics--Juvenile literature. | Children's questions and answers.
Classification: LCC QC360 .H345 2016 (print) | LCC QC360 (ebook) | DDC 535--dc23
LC record available at https://lccn.loc.gov/2016051371

Printed in Brainerd, Minnesota, United States
1 2 3 4 5 6 7 8 9 0 21 20 19 18 17

052017
150517

Photo Credits
Every reasonable effort has been made to trace ownership and to obtain permission to reprint copyright material. The publisher would be pleased to have any errors or omissions brought to its attention so that they may be corrected in subsequent printings. The publisher acknowledges Getty Images, Alamy, iStock, Dreamstime, and Shutterstock as its primary image suppliers for this title.